Work and Simple Machines

Lesson 1
How Do Levers Help You Do Work? 2

Lesson 2
What Are Inclined Planes? 10

SCHOOL PUBLISHERS

Orlando   Austin   New York   San Diego   Toronto   London

Visit *The Learning Site!*
www.harcourtschool.com

Lesson 1

How Do Levers Help You Do Work?

VOCABULARY
work
lever
fulcrum
pulley
wheel-and-axle

You do **work** when you use force to move an object in the direction of the force.

A **lever** is an arm or rod that turns around a fixed point. The **fulcrum** is the fixed point.

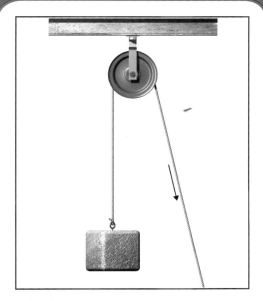

A **pulley** is a grooved wheel that has a rope or chain around it.

A **wheel-and-axle** is made up of a large wheel and a small wheel that turn together.

READING FOCUS SKILL

CAUSE AND EFFECT

The **cause** is what makes something happen. The **effect** is what happens.

Look for examples of **cause** and **effect** as you read about simple machines making work easier.

Work

Work is one of those words that has an everyday meaning and a scientific meaning. You probably think you are working right now. You are reading this book. But a scientist would say you are not working. A scientist would say that **work** occurs only when a force moves an object in the direction of the force. You're doing work when you ride your bike or kick a ball.

Suppose you try to move a heavy chair. You push and pull on it all day long. But it doesn't move. You haven't done any work. But if you pick up a feather, you've done work.

You can find out how much work you do. You multiply the force you use by the distance the object moves. The formula is *work = force x distance*. Force is measured in a unit called newtons (N). Distance is usually measured in meters. The unit for work is called the joule.

How will you know if this boy does work? ▶

People do work. Machines do work too. Machines make work easier. There are two kinds of machines.

Simple machines are all around us. They have only a few parts. They move only one way. Simple machines include levers, pulleys, and wheel-and-axles.

Two or more simple machines make up a *compound machine.* These machines can have many parts. A car is a compound machine.

Machines don't reduce the amount of work you need to do. They change the kind of work or they change the direction of work. They make the work easier.

 Tell what the effect of work is.

▼ These are compound machines.

5

Levers

Levers are one type of simple machine. Levers are all around you. They are in scissors, rakes, can openers, shovels, and seesaws. A **lever** is an arm or rod that turns around a fixed point when you apply a force. Parts of levers have special names.

Think about a seesaw. One child sits on one end of the seesaw. She applies the *effort force.* She sits on the effort arm. The *effort arm* of a lever is where the force is applied. Her weight moves the seesaw down.

The seesaw is propped on a center point. This is the **fulcrum**, or fixed point. The seesaw moves on this fulcrum. When the girl sat down on her end of the seesaw, the other end moved up.

The end that moves is called the *resistance arm.* The resistance arm applies a *resistance force.* If another child is sitting on that end, that child is the load. The load moves up with the resistance arm.

A lever applies a force over a distance. People use levers to lift heavy objects. If a lever is long and strong enough, you can use it to lift a car.

 Explain how a lever lifts a load.

The resistance force on the wheelbarrow is between the effort force and the fulcrum. ▶

Pulleys

Pulleys are another common kind of simple machine. You use a pulley when you raise window blinds. You also use a pulley when you raise a flag up a flagpole. People often use pulleys to lift very heavy or very large objects.

A **pulley** is a grooved wheel that has a chain or rope around it. When you pull the chain or rope down, the pulley turns. The rope or chain moves up and down.

Pulleys are related to levers. The fulcrum on a pulley is in the center of the wheel. The pulley's rope applies the effort force. The effort arm and the resistance arm are on opposite sides of the pulley wheel. The load being lifted is the resistance force.

Pulleys make work easier because it's easier to pull down on something than to pull up.

 Tell how a pulley can raise a flag.

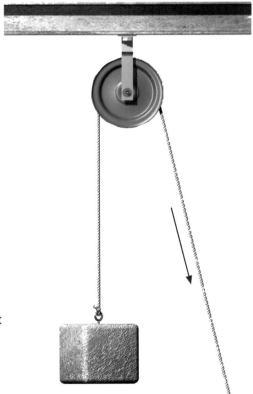

A pulley changes the direction of the effort force but not the size of the effort force. ▶

7

Wheel-and-Axles

A **wheel-and-axle** is another simple machine. It is a large wheel and a small wheel that turn together. Doorknobs, bicycle wheels, and car steering wheels are all examples of a wheel-and-axle. The smaller wheel is the axle. It is often a simple rod.

A wheel-and-axle makes work easier because you apply less effort over a larger distance on the wheel. You would have to apply more force to move the axle itself.

For example, a driver turns the steering wheel of a car. The wheel turns a greater distance than the axle. But the driver applies a smaller force to move the wheel. The driver would have to apply a much greater force to move the axle directly.

Some wheel-and-axles increase the distance through which a force acts. In a Ferris wheel, an effort force turns the axle a small distance. The Ferris wheel moves faster and farther than the axle.

 Explain how wheel-and-axles make work easier.

A sailboat crank is a wheel-and-axle. The handle acts as a large wheel. It turns a small wheel. This wheel lets out the sail. ▼

More About Machines

Machines give us an advantage. That's why we use them. We use the term *mechanical advantage* to describe how much a force is increased by using the machine. The greater the mechanical advantage, the less force you must use to do work. The smaller the mechanical advantage of a machine, the more force you must use to do work.

Efficiency is the amount of work a machine can do compared with how much work you have to put into the machine. No machine has an efficiency of 100 percent. Efficiency is always reduced by forces like wind and friction. Your bicycle's efficiency is lessened by wind resistance on you and the friction between the tires and the road.

 Name a force that lessens the efficiency of this kayak.

The oar is a lever. The water pressing against the oar is the resistance force. ▶

Complete these cause and effect statements.

1. Work causes an object to _____ in the direction of the force.

2. A force on the effort arm of a lever makes the _____ arm move.

3. You apply less effort over a greater distance on a wheel to make an _____ move.

4. Friction causes efficiency to be _____ .

Lesson 2
What Are Inclined Planes?

VOCABULARY
inclined plane
wedge
screw

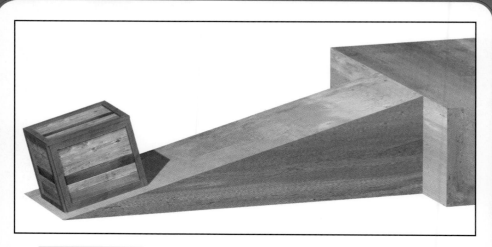

An **inclined plane** is a simple machine. It is any flat, sloping surface.

A **wedge** is another simple machine. It is made of two inclined planes placed back to back.

It doesn't look like an inclined plane, but a **screw** is actually an inclined plane wrapped around a cylinder or a cone.

READING FOCUS SKILL
CAUSE AND EFFECT

The **cause** is what makes something happen. The **effect** is what happens.

Look for **cause**-and-**effect** relationships in the use of inclined planes and wedges.

Inclined Planes

The inclined plane is the simplest of all the simple machines. It has only one part. An **inclined plane** is a flat, sloping surface. Another name for it is *ramp*.

Inclined planes are used mostly to lift or lower objects. Roads leading up a hill or mountain are inclined planes. So are stairs. Movers use ramps to put furniture on trucks. You have probably seen ramps on sidewalks for people in wheelchairs to use.

An inclined plane can change the amount or direction of the effort force. Like other simple machines, an inclined plane doesn't reduce the amount of work needed to do a job. But you can use less force. This is because the distance is longer. You lift by using a smaller force over a longer distance.

▼ This long ramp is an inclined plane.

Inclined planes are used to help you lift heavy loads. It is easier to move something up a ramp than to pick it straight up.

Look at the pictures below. Suppose you need to get the box up to the platform. In the first picture, you don't see a ramp. The only thing you can do is pick up the box. If it is heavy, you would have to apply a lot of force. You also might hurt your back.

The next picture shows a steep ramp. The ramp isn't very long. If you push the box up this ramp, it will take a lot of force. But not as much as in the first picture.

The next picture shows the best way to move the box. The ramp is longer. You use less force over a longer distance.

Inclined planes are also used to slow things down. In mountain areas, you may see gravel ramps at the side of the road. They go up the mountain a short distance. This is for trucks that have been going downhill. Drivers can lose control of the large truck. So they drive up this ramp. The ramp slows them down. The driver can then get control.

 How does using a longer inclined plane affect the force needed to lift something?

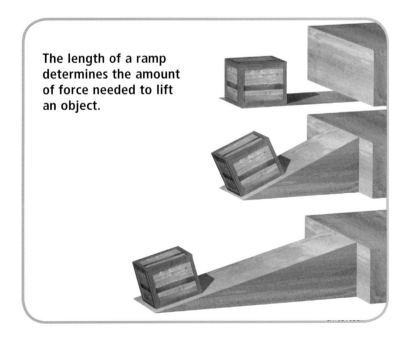

The length of a ramp determines the amount of force needed to lift an object.

13

Wedges

A **wedge** is a type of inclined plane. It can have one or two sloping sides. Most have two. Wedges have one thick end and one sharper, narrower end. They are used for splitting or separating things. A knife, an ax, a plow, and a nail are all wedges. You apply the effort force to the thick end of a wedge. The inclined planes increase this force. They also change the direction of the force. The inclined planes make the force go outward. The material separates as the wedge moves through it.

Wedges can also lift or push objects. A doorstop is a wedge. You put the sharp end of the wedge under the door. The wedge pushes upward on the door. This force raises it slightly. The door applies an equal and opposite force on the doorstop. This pushes the doorstop against the floor. Friction holds the doorstop in place.

 Tell how the shape of a wedge causes materials like wood to split apart.

A knife is made of two inclined planes joined as a wedge. The sharper the edge, the less effort force you need to use. ▼

Screws

A **screw** is an inclined plane wrapped around a cylinder or cone. To use a screw, you apply less force over a greater distance. A screw also changes the direction of force. As a screw is turned, it moves up or down.

The pictures below show how screws have ridges, or *threads*. These threads operate like tiny inclined planes. Screws with more threads per centimeter have gentler inclined planes. They require less force to turn. Screws with fewer threads per centimeter have steeper inclined planes. They require more force to turn.

 Explain why it's easier to use a screw with more threads per centimeter.

A wood screw acts as a long, spiraling wedge.

Review

Complete these cause and effect statements.

1. Increasing the length of a ramp makes it _____ to push a box up it.

2. Effort force applied to a _____ makes the force go outward.

3. Screws with fewer threads per centimeter have steeper inclined planes, and require more _____ to turn the screw.

4. As you turn a screw, it moves _____ or _____.

GLOSSARY

fulcrum (FUHL•kruhm) The fixed point that the effort arm of a lever moves around.

inclined plane (in•KLYND PLAYN) Any flat, sloping surface.

lever (LEV•er) An arm or rod that turns around a fixed point.

pulley (PUHL•ee) A grooved wheel that has a rope or a chain around it.

screw (SKROO) An inclined plane wrapped around a cylinder or a cone.

wedge (WEJ) A simple machine made of two inclined planes placed back to back.

wheel-and-axle (weel•and•AK•suhl) A large wheel and a small wheel that turn together.

work (WERK) That which is done on an object when a force causes the object to move in the direction of the force.